Contents

Introduction

Tropical plants have a most important role to play in an aquarium. Apart from being chosen for their decorative leaves and colour, they play an essential part in the lives of the fish. During daylight hours, the plants absorb carbon dioxide from the water where it is released by the respiration of the fish, while also giving off valuable oxygen. At night it is in reverse, the plants take in oxygen and lose the carbon dioxide – at this time the fish do not require too much oxygen, as they are usually resting and inactive.

The aim of this guide is to give a very basic outline of the variety of aquatic plants suitable for the aquarium, and the conditions in which they require to be grown.

The aquarium for many years has been recognised as having a relaxing effect on people with stress and tension; the pleasure of watching the plants grow and flourish is very satisfying, and hopefully this will be achieved.

An attractive aquarium using bog wood to add to the display.

HAMLYN
Pet Guides

TROPICAL
AQUARIUM
PLANTS

Patricia Horeman

HAMLYN

Published 1985 by Hamlyn Publishing,
Bridge House, London Road, Twickenham, Middlesex

Copyright © Hamlyn Publishing 1985
a division of The Hamlyn Publishing Group Ltd

All rights reserved. No part of this publication may be reproduced, stored in
a retrieval system, or transmitted, in any form or by any means, electronic,
mechanical, photocopying, recording or otherwise, without the prior
permission of Hamlyn Publishing.

ISBN 0 600 30671 2
Printed in Italy

*Some of the illustrations in this book are reproduced from other books published
by The Hamlyn Publishing Group*

Planting Medium

The endeavour is to supply, as near as possible, a medium that is suitable for all types of plants, i.e. submersed plants and bog type plants with rhizomes or bulbs.

The ideal medium is 1 or 2-mm natural gravel, lime free where possible, with a depth of at least 5–6.5 cm (2.5–3 in); a layer of unwashed gravel (one third) topped with the remainder of washed gravel (two thirds), washed meaning being in running clear water.

Aquatic plants do not rely solely on receiving nutrition through their roots; they are also capable of absorbing nutrients through their leaves.

It is not advisable to put peat or loam underneath the gravel as the water becomes polluted and high in nitrates and encourages the growth of blue/green algae which chokes the plants and is highly toxic to fish.

A satisfactory growth of plant can be obtained in time when the nutrients become richer due to the accumulation of fish detritus and debris.

This set up uses only plants to provide both depth and cover.

Lighting

Artificial light is essential to the growth of aquatic plants. It is also important when positioning the aquarium in a room. Ideally this should be away from a window, as excessive light will increase an algae problem.

In the temperate zones, the length of day varies with the seasons. In the tropics, the length of day and night are equal, so it is necessary to lengthen the day in the winter months in the temperate zones, with the addition of artificial light for at least 10–12 hours a day. The length of time the light is on is more important than the intensity of light.

Plants respond best to light that is in the red and blue zone of the spectrum. Special fluorescent tubes are available with this light called 'gro-lux'. Equally effective are clear tungsten bulbs.

Another type of lamp that is popular for the growth of aquatic plants is a mercury vapour lamp, and this is usually suspended over the aquarium. The lamp consumes about 80 watts but can transmit light up to the value of 120 watts. House plants also benefit from this light.

An aquarium cover housing the standard type of lighting.

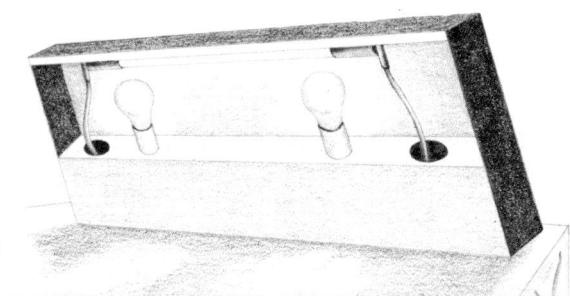

Water

Fortunately, aquatic plants do not need demanding water conditions to grow and increase. In fact the conditions we can provide for them in the aquarium are often far superior to those in their natural habitat.

The majority of plants are happy with a water hardness up to 12–15 German degrees (°DH). It is only the very delicate plants, like *Aponogeton fenestralis* (Lace Plant), that prefer a softer water. The PH values (acidity and alkalinity) are also not particularly critical – once an aquarium becomes established and the fish are introduced, it becomes very obvious as to which plants are going to succeed and those which will fall by the wayside.

The type of aquarium decoration chosen also contributes towards the quality and type of water. For example, natural bog wood will increase the acidity in the water, whereas Westmorland rock, which is limestone, will tend to make the water alkaline. Either choice of decorations is not detrimental to fish or plants.

Temperature

As with the water conditions, the temperature range is not too critical. An average temperature of about 22–28°C (71–82°F) is very suitable for most of the hardier types of plants and also meets the general conditions required for the tropical fish. An occasional drop of a degree or two is not harmful.

Also the installation of an under-gravel filter (see Filtration) assists in obtaining an overall temperature, so there is no area that is colder than another. In some cases,

This attractive type of aquarium is very popular in the home as all the equipment is hidden behind wooden panels.

where the plant is grown on a bulb or corm, they need a resting period where the temperature needs to be cooler for a short period.

The pH scale, showing the colour variations that are obtained when using the indicator dyes provided in a pH kit.

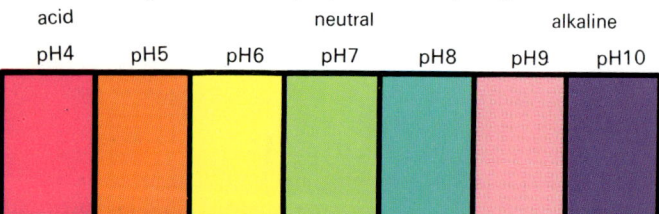

acid   neutral   alkaline

pH4   pH5   pH6   pH7   pH8   pH9   pH10

Filtration

There have been many controversial points of view concerning the virtues of biological filters (under-gravel) and their influence on the growth of plants. There are several reasons in their favour which will help to settle any doubts.

The biological filter works on the principle of absorbing the water through the gravel to the filter plates, or tubes, at the base of the aquarium. The detritus is assimilated by the beneficial aerobic bacteria which are naturally present in the gravel. This in turn brings all the

An under-gravel or biological filter. The arrows show the direction of the water flow.

nutrients from the water down to the plant roots where it is needed. Also the temperature of the water is even throughout the aquarium with no cold spots, and the warm water is drawn down to the roots of the plant so that they have 'warm feet'. The biological filter is giving oxygen and filtering at the same time producing excellent clarity of water.

When the gravel has loam or peat underneath, there is no air circulation and it becomes airtight, the bad bacteria forms (anaerobics) in the gravel and the plant roots become black and rot.

Other types of mechanical filters are also excellent in taking out the suspension in the water leaving it crystal clear. Also the motorized filters have facilities to add filter mediums to improve the quality of water to your requirements.

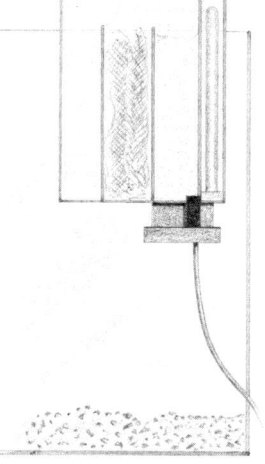

Right: external box filter with electric water pump.
Below: non-submersible electric water pump for the top of a biological filter's lift tube.

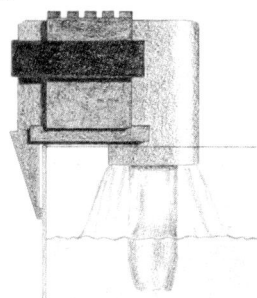

Common Problems

The most common problems encountered are lack of growth, thinning off of plants, and falling leaves. This is usually attributed to poor lighting, as explained in the Lighting chapter. Most tropical plants are imported from the tropical zones and have an equal length of day. If the daylight hours are extended to a minimum of 10 hours, the plants will show signs of improving.

In a newly set-up aquarium, the plants need time to establish themselves and, once the fish are introduced, the natural debris will provide the nutrients for them to survive initially, while the addition of some aquarium fertilizer will also give them a start.

With the larger leaf plants, e.g. Amazon Swords, Cryptocorynes and the larger Hygrophila species, there may be some yellowing of the leaves, brown spots or rotting of the roots. These particular plants are mostly cultivated out of water (emerse) or imported from Asia, where the humidity is high; plant leaves grown in these conditions become stronger and thicker. When they are introduced into the aquarium, the plants become completely submersed and the emerse leaves start to rot and disintegrate. When this happens it is best to remove these leaves allowing new submersed leaves to grow.

Some aquatic plants offered for sale that develop rotting leaves and stems may not be true aquatics and are in fact more happy as house plants. These plants have a limited life but there are some exceptions to the rule that can take on an aquatic role and survive.

Snails have always been advocated as useful for aquariums as a cleaning agent, but they are enemies of the plants, especially the round flat type called Red Ramshorns (*Plānorbis corneus*); they enjoy eating the leaves, giving the plants a moth-eaten appearance. The only snail which does not have a bad reputation is the Malayan Snail (*Melania tuberculata*). They enjoy turning over the gravel looking for food; also they are live bearing, so now and again colonies of baby snails are seen floating on the surface of the tank – the population must be kept in check for they would overrun the aquarium and be unsightly.

Three types of snail found in an aquarium: 1 Red Ramshorn; 2 *Bullinus*; 3 Malayan Snail.

Snails can be good indicators that something is starting to go wrong, e.g. the water is becoming polluted. If the snails start to move up the glass of the aquarium to the surface, this can mean either the water needs changing, or the gravel has gone bad and needs renewing.

Another enemy of the plants is the vegetarian fish with a healthy appetite for plants – these fishes are of course to be avoided, if possible, in aquariums where plants are to grow. Among these species of vegetarian fish is the Silver Dollar (*Metynnis roosevelti*), Tin-foil Barbs (*Barbus schwanenfeldi*), *Leporinus* species, *Distichodus* species, and scats (*Scatophagus*). However, some vegetarian fishes are useful in cleaning up any excess algae deposits on leaves, e.g. *Gyrinocheilus* (the Sucking Loach), *Otocinclus*, *Plecostomus*.

In the section on lighting, it is mentioned that an aquarium is best situated away from a window because it could cause an algae problem. Algae is a lower form of plant life and there are several different forms. They thrive in light, and as the days lengthen and extra light falls on the aquarium, the algae will form a green growth on the glass and will attach itself to the plants and choke them. Also, if the aquarium lighting is left on too long, algae will form.

Algae can be kept in check either by cutting down the light source, by introducing algae-eating fishes, or by the use of one of the many chemical algicides which are available. The algae growth on the glass can be removed with a scraper; the most effective kinds are a plastic (or metal) stick with a razor blade on one end and a 'V' on the other end (to use as a planting stick) or an algae magnet which saves hands from getting wet.

Selecting Plants and Planting

This is a most pleasant task as there is such a large variety to choose from. The basic idea is to form as naturally as possible an underwater garden, blending colours and leaf forms to give a backdrop for the fishes and contrast to the rocks and various decor chosen, but also allowing swimming room for the fishes.

When purchasing plants, make sure that they look fresh and healthy and have a good colour; prior to planting, remove any dead leaves and check for any snails eggs in the form of a jelly attached to the leaves. Keep the plants from drying out until ready to plant in the aquarium.

The first step is to arrange the rocks or wood, etc., then start with the tall background plants such as Vallisneria, Cabomba, Densa, and Myriophyllum. These are good

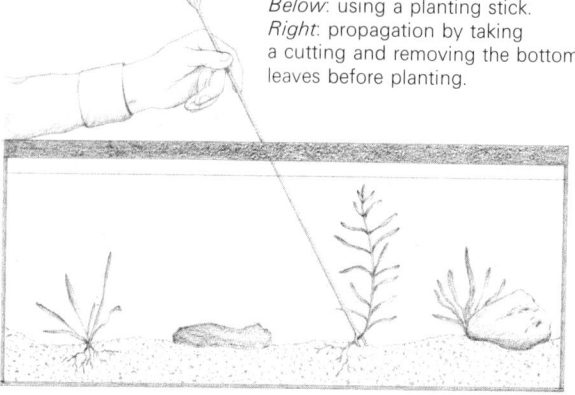

Below: using a planting stick.
Right: propagation by taking
a cutting and removing the bottom
leaves before planting.

plants to hide airlifts and heaters in the aquarium. The medium height and bushy plants like Wisteria, Ambulia, and Hygrophila, can then be put round the sides. The dwarf plants, Acorus, Pygmy Sword plants, and small Cryptocorynes are saved for the foreground, bearing in mind that some room must be left for growing.

It is easier to plant an aquarium when it is half full of water. When planting singular rooted plants, it is best to ease them up a little bit after planting, so the crown of the plant is not buried, as this tends to make the plant rot. Also save a space for a centre plant, e.g. an Amazon plant or Aponogeton. Top up the aquarium taking care not to disturb the plants.

Once the lighting and the heating are installed, it should only take approximately 7–10 days for the plants to start to establish themselves.

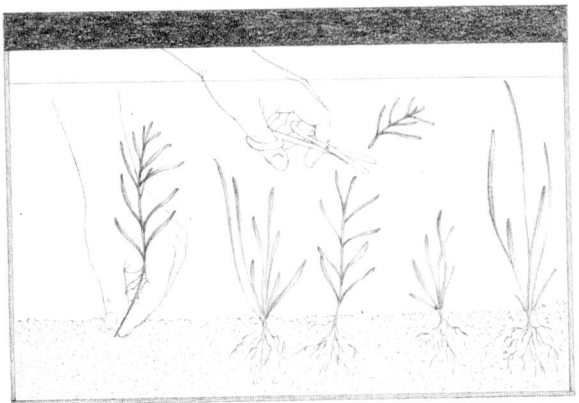

The Most Popular Aquarium Plants

Aponogetons are most attractive and are grown from corms. The majority are very hardy. The most commonly found are the species from Sri Lanka while the rarer most beautiful specimens are from Madagascar. A great many Aponogetons throw up a flower spike out of the water, sometimes forming ripe seeds. The species from Sri Lanka have one flowering spike, while the species from Madagascar always have at least two flowering spikes.

They have a resting period in early autumn when the water temperature should be cooler, with the warmer temperatures in the spring, when they will start to develop new leaves. Often found sold in bulb form.

Aponogeton crispus. From Sri Lanka, it is the easiest to grow; bright green nearly transparent leaves, with heavily crinkled edges hence 'crispus'. Grows up to 20–25 cm (8–10 in) long, and prefers a bright light.

Aponogeton fenestralis (Lace Plant). From Madagascar, it is not the easiest plant to grow because of its very delicate lacy leaves. A shaded position is ideal with soft and slightly acid water.

Aponogeton stachyosporus (Undulatus). A very interesting species, it is viviparous (live-bearing) and produces small bulblets where the young plants form. It is undemanding, requires adequate light; the leaves are wavy, semi-transparent, light green (dark green in places) and up to 40 cm (16 in) long.

Bacopa amplexicaulis. An amphibious plant which adapts well to aquarium culture. Amplexicaulis is the larger-leaved variety of the two *Bacopa* species

Top left: **Aponogeton fenestralis**
Above: **Barclaya longifolia**.
Left: **Aponogeton crispus**.

mentioned. Best planted in groups; cuttings can be taken by nipping between the leaf axil and replanting. Prefers shade.

17

Bacopa monnieri (Baby's Tears). Small round leaves – slightly stiff, and light green, without veins. Conditions and cuttings as above but happier with more light.

Barclaya longifolia. Species belonging to the water lily family originally from Thailand. Grown from a corm, it has very delicate fragile leaves, is soft green in colour with a light pink underside; leaves are long and tapered at the top – 30 cm (12 in) long and 25–40 mm wide. Temperature is important – up to 28°C (82°F) if plant is to flower (appears on the water's surface) and not lower than 20°C (68°F). Easily attacked by snails.

Cabomba aquatica (Fan Wort). These are completely submersed plants; the very fine leaves are divided into segments and are very decorative and feathery. There are several species ranging from a dark green to light green. Prefers bright light especially in the winter months; can be increased by cuttings.

Ceratophyllum demersum (Hornwort). Mostly found in cold water aquariums where it grows more densely and a darker green. The leaves are fine and brittle resembling a pine cone. In tropical waters it tends to get drawn and thin but can look attractive.

Ceratopteris thalictroides (Indian Fern). This is a fern-type plant, light green and quite brittle. Prefers soft acid water, it is best to save some young plantlets to continue for the spring as the old plants tend to die back in the winter. Guppy breeders advocate having this plant with the fish as they go well together.

Ceratopteris cornuta. This is the floating variety of the above species and, being of a fern-like family, develops young plantlets on the leaf edges. The fine roots offer safe cover for young fishes to hide in.

18

Cryptocorynes

The Crypotocoryne family is amongst the most popular found in the aquarium; all the plants are originally from Asia. In their natural habitat, they grow partly as a bog plant. This is the only time when they will flower out of water.

There are up to 100 species of Cryptocoryne which have been named and re-named by botanists. The most well known are briefly described here. They also have a tendency to make the water acid so can be an ideal companion for the spawning of some species of fish. Most will tolerate medium/hard water and a temperature of 20–28°C (68–82°F).

Cryptocoryne affinis (*C. haerteliniana*). One of the oldest favourites in the aquarium with dark blue/green leaves and claret colouring underneath; grows and multiplies quite quickly as a rhizome – can be split up and divided if the clump becomes too dense. Leaves up to 30 cm (12 in) long.

Cryptocoryne becketii. Comes from Sri Lanka and is the most hardiest variety, suitable for all aquarium conditions. Medium-sized leaves, 15 cm (6 in) long; olive green on the surface, light pink underneath.

Cryptocoryne blassi. Originally named after Herr Blass of Munich and revised to *C. siamensis*. From Thailand, it is a large plant, with leaves up to 20–25 cm (8–10 in) in length – dark green on the surface, deep wine colour underneath. It is one of the Cryptocorynes that will tolerate hard water.

Cryptocoryne ciliata. There are two varieties of this Cryptocoryne. The one most often found is the small variety 15–25 cm (6–10 in) long with light green, slightly fleshy spear-shaped leaves. The larger variety is found in the Malay Peninsula and grows up to 1 m (3 ft) in height! The small variety is quite hardy, and is most suitable for the aquarium.

Cryptocoryne nevillii. Neat little plant from Sri Lanka – grass like, with quite stiff leaves from 5–10 cm (2–4 in) long. Often used for planting in the foreground of aquariums. Multiplies quite quickly on runners, and makes a nice dense clump.

Echinodorus (Amazon Swords)
These are a larger-sized aquarium plant, usually used as a centrepiece, found in tropical America. In their natural

Below left: **Cryptocoryne affinis**.
Below right: **C. nevillii**.

Echinodorus paniculatus.

habitat they grow out of water (emerse) – they are cultivated mostly emerse for commercial growing in Asia and Europe. They adapt to their submersed conditions very well, losing their outer leaves and producing new underwater leaves. There are some smaller species that are popular which propagate by sending out runners with plantlets on, sometimes known as 'Pygmy Chain Swords'. Essential to have good light; water conditions not critical.

Echinodorus bleheri (Paniculatus). One of the most well-known Amazon Sword Plants. Good tempered and easy to grow, the leaves are quite broad about 40–60 mm (1.5–2.3 in) and grows to a height of 20–40 cm (8–16 in). The plant develops up to at least 20 leaves and throws off a runner, with plantlets attached, which can be separated from the parent plant when they become large enough.

Echinodorus horemanii. A recent introduction to the aquarium world. Only grows submersed which is unusual for an Echinodorus. Beautiful, glossy dark green, strap-like leaves growing up to 40 cm (16 in) long and more; it can withstand a wide temperature range from 14°–27°C (57°–81°F). An ideal centrepiece, very decorative.

Echinodorus martii (Ruffled Sword). Another sought after specimen for a centrepiece. Grows from a rhizome; long, light green leaves which are very wavy and undulating, hence 'Ruffled Sword'. If given plenty of light, it will produce excellent leaf formation and will be a credit to the aquarium.

Echinodorus osiris. Beautiful plant that has two types of leaf colours: the young leaves are red/brown and grow into large, wide, light to medium green leaves. Sometimes in cooler temperatures – i.e. 20°C (68°F) – the leaves become a more intense red.

Echinodorus parviflora (Black Amazon). From Peru, this medium-sized Amazon Plant has spear-shaped leaves up to 20 cm (8 in) long with almost black vein markings on the leaves – hence 'Black Amazon'. It can produce up to 40 leaves all symmetrically arranged, which makes it a most attractive plant.

Echinodorus tenellus (Pygmy-chain). A delightful miniature sword plant growing only about 5 cm (2 in) high. The smallest plant in this group (grass-like in appearance), it propagates by root runners. Under good light conditions, it can produce a 'carpet' of green foliage. Ideal as a foreground plant.

Single-stemmed Plants

This next group of plants are mainly singular-stemmed, i.e. one can take cuttings and so multiply the specimens and make a lush, bushy appearance in the aquarium.

Egeria densa. This is one of the most used background plants and is very adaptable to many conditions. Cuttings can be taken and the plant kept to a height to suit the individual requirements. It throws out roots along the stems and can anchor itself into the gravel. It is dark green in colour with short grass-like leaves all the way up the stems ending in a whorl of leaves at the top. It has a unique property – it can absorb the lime in hard water therefore making it softer.

Hygrophila corymbosa (*Nomaphila stricta*). Commonly called Giant Hygrophila. Leaves look like Stinging Nettles, spear-shaped with serrated edge. Can grow out of the top of the aquarium where it will form small blue flowers. Best kept trimmed by cuttings, nipped off between the leaf axil; replant the cutting. If through lack of light the plant becomes 'leggy', nip off the top leaves and replant.

Hygrophila difformis (Wisteria). An attractive fern-like plant. Very fast growing and easily propagated by stem cuttings, it has beautiful light-green finely-cut leaves and makes a good bushy plant. It does not require too bright a light – at a temperature of about 25°C (77°F) the leaves will keep their finely-cut appearance.

Hygrophila polysperma. A most undemanding plant which thrives in normal aquarium conditions. The stem is quite delicate and brittle but will tolerate hard water. Cuttings can be taken to increase them; needs plenty of light.

Limnophila sessiflora (Ambulia). Similar to Cabomba in appearance but more feathery and lighter green; the stems are quite fleshy and cuttings can be taken. Grows quite fast but prefers more acid water. The fine whorls of leaves make excellent cover for baby fish.

Ludwigia repens (Natans). A pretty stemmed plant with small oval leaves growing up the stems – leaves have a dark green surface with pink to dark red underneath. It is very versatile, growing equally well in a cold water or tropical aquarium.

Microsorium pteropus (Java Fern). A fascinating fern plant, which in its natural habitat grows as an amphibious plant. A stiff pointed oval-shaped leaf, quite rough in texture growing along a rhizome. Looks most attractive attached to a bog-wood decoration or cork

Below left: **Echinodorus martii**. *Below right:* **E. tenellus**.

where it will anchor itself quite happily. Young plantlets form on the underside of the leaf from spores. As with all land ferns these can be detached when large enough.

Myriophyllum aquaticum. A very fine feather-like plant – graceful and delicate. An ideal background plant, it does well in deep tanks, and is undemanding temperature-wise. It is important to have clear water with these types of plants – any sediment settling on the leaves tends to make them deteriorate.

Nymphaea stellata. A miniature water lily which sometimes flowers on the surface of an aquarium. Triangular-shaped leaves, dark brown to purple in colour, will send up aerial leaves to float on the surface; these can be taken off as too many visible stems can look unsightly.

Nymphoides aquatica (Banana Plant). This novelty plant with heart-shaped leaves comes from the United States. The root stock is formed like a bunch of bananas, which in fact is the food storage for the plant. The roots are not planted in the gravel, to enable the 'bananas' to be visible.

Rotala macrantha. A single-stemmed plant with soft pink leaves, almost papery in texture. A good contrasting plant against the dark greens; prefers to be in half shade, water conditions not critical.

Sagittaria subulata (Natans). A very hardy single-rooted plant with stiff grass-like leaves; adapts to any aquarium conditions. It soon spreads by root runners and makes a dense clump. Adds variety to a plant collection.

Vallisneria americana (*V. spiralis f. tortifolia*). Another favourite for the aquarium, with grass-like leaves which twist and spiral. Also produced by root runners, it is single-rooted. Oxygen bubbles can often be seen coming off the leaves to the surface.

Vallisneria spiralis. A straight, grass-like leaf variety, opposite to the *tortifolia* but equally adaptable in the aquarium. Will grow up to 45 cm (18 in) and does well if given plenty of light.

Vallisneria gigantea. As the name implies, a giant variety of the Vallisneria family. Ideal for deep aquariums, it has broad ribbon-like green leaves and does well in either tropical or cold water aquariums.

Vesicularia dubyana (Java Moss). This is an ideal moss to attach to bog-wood, or cork bark (like the Java Fern). It prefers a shady position; once it has established itself, it grows quite quickly, making the wood blend

Below left. **Hygrophila difformis**. *Below left:*
Microsorium pteropus.

naturally in the aquarium. It is an ideal spawning medium for egg-laying fish.

Floating Plants

There are several varieties of floating plants, but what follows are the most suitable for the aquarium.

Azolla caroliniana (Fairy Moss). With tiny leaves in clusters, coloured from green to rust, it will cover the surface of the aquarium. Needs to be thinned out occasionally, as it will shade the aquatic plants too much.

Riccia (Crystalwort). Forms a thick mass of small branches forked in all directions, just under the surface of the water. Frequently used for bubble nest building fishes, where the bubbles adhere to the *Riccia*.

Salvinia auriculata. Floating leaves, round to oval-shape in pairs, developing in clusters. The surface of the leaf is similar to a Cats Tongue in appearance, the tiny hairs repelling any droplets of water on their surface.

Bog or Marsh Type Plants

These plants are grown or found in mainly marshy areas, but will adapt and grow successfully in aquarium conditions. Also often used in terrariums or paludariums, where humidity is produced for the keeping of amphibians, etc.

Acorus gramineus var pusillus (Japanese Rush). Dwarf form of the Acorus family, it is grown on a rhizome similar to that of an iris. The stiff green leaves arrange themselves in a fan shape; very slow growing but keeps its colour and shape well.

Aglaonema simplex. Often referred to as a Malayan Sword, it has no connection whatsoever with the sword plant family. Grows on a rhizome with thick fleshy spear-shaped leaves which are shaded medium green. Does not increase under aquarium conditions.

Anubias nana. A rare plant from Africa, it also grows on a rhizome. This is a dwarf variety with stiff oval leaves – a glossy green on the surface with a lighter green underneath. Young plants form slowly along the rhizome. It will sometimes flower underwater. The flower looks like a miniature arum lily. Cultivated

commercially and sold in a small lattice pot with special growing medium, the pot can be submerged in the gravel so not disturbing the roots.

Ophiopogon japonicus (Fountain Plant). Not really a true aquatic plant but is quite successful in the aquarium. Thin stiff grass-like dark green leaves curving in an arch on either side like a fountain; slow growing, holds well, and is 'fish resistant'.

Spathiphyllum wallisii. A cultivated plant specially grown for aquarium conditions - it is normally a greenhouse plant growing out of water (emerse). It keeps in very good condition in the aquarium but grows very slowly. A long spear-shaped leaf, slightly wavy, dark green with clear vein markings. A centrepiece, it will grow to about 30 cm (12 in).

Far left: **Rotala macrantha**.
Left: **Vallisneria spiralis**.
Right: **Acorus gramineus var pusillus**.

29

Syngonium albolineatum (Goose-foot). Another species adapted for the aquarium; triangular shaped leaves variegated green and white, on a rhizome. Prefers acid water but not fussy with temperature.

This next small collection of plants are marsh or terrestrial plants which again have been adapted for the aquarium and are among the varieties which are now commercially grown in small lattice pots.

Chlorophytum bichetii (Spider Plant). This plant has become more recognised as an aquarium plant with its green and white striped narrow leaves, and little white bulblets for roots. These bulblets are only required when the plant is growing emerse; it is best to remove them when planting in the aquarium as they tend to rot and turn the gravel black. It seems to be quite happy in the aquarium providing the temperature is not too high.

Below left: **Spathiphyllum wallisii**. *Below right:* **Salvinia auriculata**.

Dracena sandriana. This plant has a limited life in the aquarium. A thick fleshy stem with stiff green and white leaves unfolding along the stem with a sharp pointed tip – there is a red variety which does not hold as well as the green and white variety.

Hemigraphis colorata (Waffle Plant). A striking plant with dark green crinkly leaves on the surface, scarlet underneath. Leaves branch out on a semi-woody stem; cuttings can be taken. Good colour in contrast to the plain greens.

Chamaedorea elegans (Palms). The least suitable plant for the aquarium, but a seedling palm tree does look attractive with its dark green matt leaves on stiff stems; it even looks good when it has turned brown and ageing!

Conclusion

Hopefully, the selection of plants described has wetted the appetite of the potential aquatic gardener and that the few basic tips have been some guide to how simple and pleasant aquatic plants are to grow. One can now sit back and enjoy the pleasures of one's own handiwork and join the ever increasing world of aquatic folk.

Happy growing!!

Index